BY FRANKLIN WATTS

ILLUSTRATIONS BY SAM SHIROMANI

Library of Congress Cataloging in Publication Data

Watts, Franklin.
Rice.

SUMMARY: Discusses the cultivation and harvesting of rice and its tremendous importance as a basic food for nearly half the world's population.
1. Rice—Juvenile literature. [1. Rice]
I. Shiromani, Sam. II. Title.
SB191.R5W4 633'.18 76-39933
ISBN 0-516-03686-6


Printed in the United States of America.
1 2 3 4 5 6 7 8 9 10 11 12 R 79 78 77

This Chinese family is eating its evening meal. Look at all the good things on the table. There are pieces of pork and chicken . . . lots of green vegetables . . . sweet and sour sauces to dip the food in . . . and bowls of boiled and fried rice.

This American family is eating its main meal. There is crispy fried chicken . . . corn on the cob . . . a big salad . . . and a loaf of bread.

Do you know what is the most important food on each table? For the Chinese family it is rice. For the American family it is bread made from wheat.

Everyone needs meat and vegetables to stay healthy. But rice and bread are the most important foods because they are *basic foods.* Each family eats these foods every single day. In very poor countries, some people eat little else but rice or bread. Without these basic foods, they would starve.

FAO Photo by F. Botts

Both bread and rice come from *cereal plants.* Cereal plants produce grains (or seeds), which can be eaten. Grains of wheat are ground into flour to make bread. Grains of rice are polished until they are creamy white. Then they are cooked until they become soft and fluffy.

There are other cereal plants. The grains of corn, rye, barley, and oats can all be made into foods. But rice and wheat are the most important. They are eaten by more people than any other cereals.

Chinese people eat rice two or three times a day. So do people in other Eastern lands.

How many times have you eaten rice this week? Did you eat a breakfast cereal made from rice? Or some boiled rice with fish, or creamed chicken, for supper? Maybe you had rice pudding for dessert.

No matter how much rice you had this week, you probably ate more bread than rice.

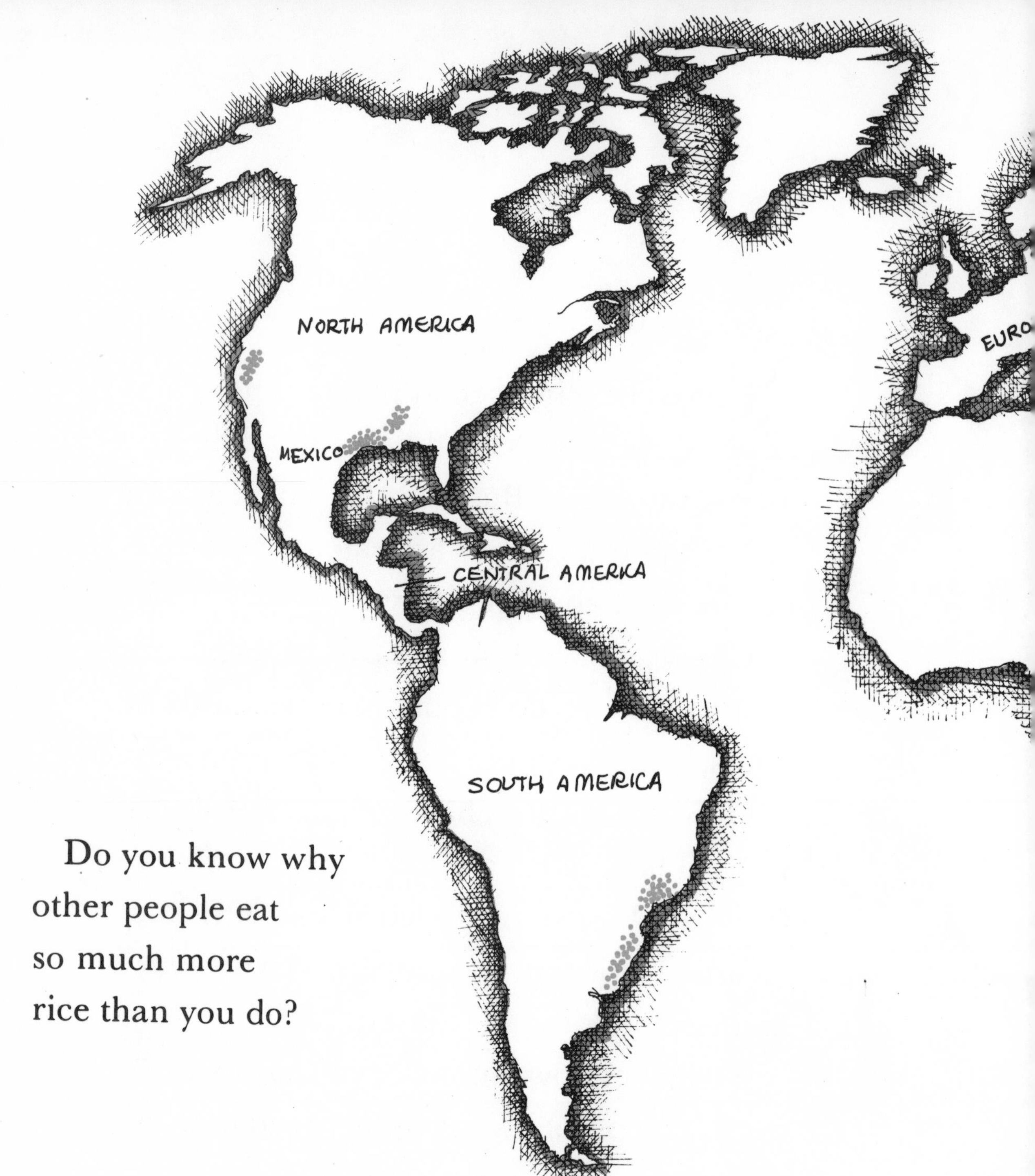


Do you know why other people eat so much more rice than you do?

Look at this map and you will see the answer. Rice grows best in hot countries that have heavy rainfall—lands such as China, India, South America, and the countries of

Southeast Asia. Wheat grows best in cooler lands. Because of the weather, rice is the most important basic food for about half of the world. For the cooler half of the world, wheat is the most important basic food.

If you went to China, or any of the rice-growing countries, you would see many rice paddies. The first thing you would notice is water flooding the rice fields. Most kinds of rice will grow only if the seeds are planted underwater.

Rice needs water. So most rice paddies are built near rivers or along coasts so that extra water can be let into the fields. Some farmers get only the water from the heavy rainfall (called *monsoons*) to flood their fields.

FAO Photo by I.W. Kelton

*Rice paddy in Burma*

*Workers in Shansi Province, China*

FAO Photo by H. Henle

The second thing you would notice about a rice paddy is the number of people working there. Planting and harvesting rice takes a lot of hard work. Most of the men and women (and sometimes the children) in the area work in the fields.

Before each season's rice is planted, the paddies have to be ploughed. Asian farmers usually have a team of water buffalo to pull the wooden plough across the muddy fields.

The water buffalo and the American buffalo are both members of the ox family. The water buffalo is widely used in rice-growing countries in Asia and Africa. It can work well in water. Right now the water buffalo is probably the most used farm work animal in the world.

*Indonesian water buffalo ploughing the fields*

FAO Photo by H. Null

*Transplanting rice seedlings*

FAO Photo by F. Botts

The rice seeds are first planted in soft muddy *seed beds.* The seeds are then covered with water.

When pale green shoots sprout from the seeds, the plants are pulled up. Workers replant the shoots of rice in flooded paddy fields. The shoots are planted in long, straight rows. This makes the job of weeding easier.

*Fertilizing the rice.* FAO Photo by F. Botts

When the heavy rains are just about over, workers move up and down the rows of plants. They pull up all the weeds by hand.

Sometimes fertilizer is spread over the fields at this time. Fertilizer helps the plants to grow tall and healthy.

Rice Council of America

At the end of the rainy season, the plants have clusters of grain hanging from the stalks. Each grain has an outer covering called a *husk*. The husk protects the grain. Husks turn a rich golden color in the sun. That means the rice is ready to be harvested.

The stalks are cut from the plants with knives or sickles, which are long curved knives. Then the stalks are tied in bundles and left to dry. The heat of the sun quickly turns the stalks to straw.

After a few days the grain has to be taken away from the straw. This is called *threshing.* Sometimes threshing is done by beating the straws with sticks, or beating them against woven mats. Some workers thresh the grain by walking around on the straw with their bare feet. Other farmers let their work animals walk on the straw.

Now the grain is removed from the straw but it is not ready for market. The rice grains are full of bits of dirt and straw. The farmer has to clean the grain.

The grain is thrown into the air and caught in a large shallow basket. It is thrown in the air and caught over and over again. This is called *winnowing.* Finally, only the heavy grains of rice are left in the bottom of the basket. All the bits of straw and dirt have been tossed away.

FAO Photo by T. Loftas

FAO Photo by S. Bunnag

*Top:*
*Camels thresh grain in India.*

*Above left:*
*Indian farmer winnowing rice.*

*Right:*
*Cambodian woman winnowing rice.*

FAO Photo by S. Bunnag

Now the rice is ready to go to a nearby mill where the husks will be taken from the grain. Some farmers use buffalo, or carts, or bicycles to carry their rice to the mill. Some Asian farmers have to carry the rice themselves. They carry it on a special kind of pole called an *A frame.*

FAO Photo by Banoun/Caracciolo

*United Nations programs bring modern ways to underdeveloped nations.*

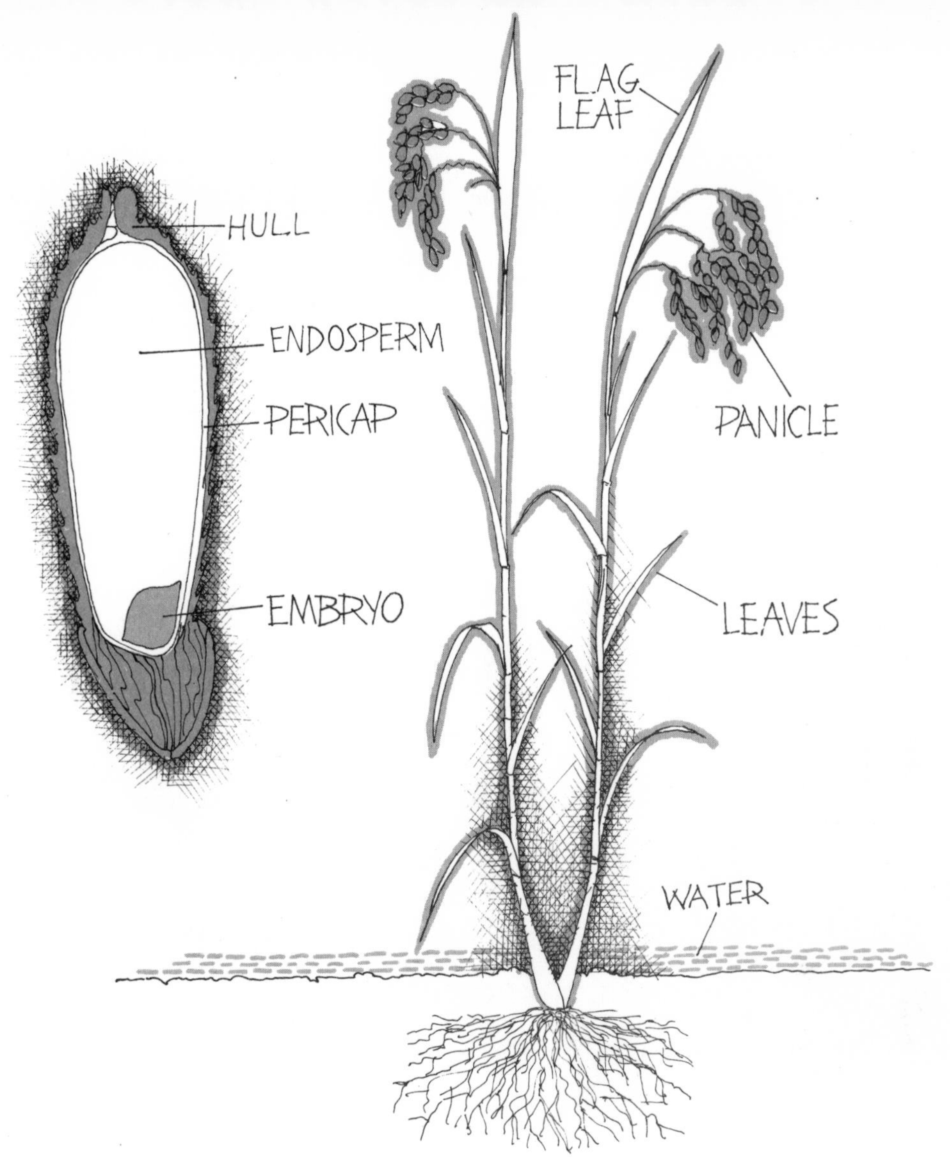


The rice you eat probably is white. But Asian people often eat rice that is still covered with a brownish coating. The coating is called *bran.* Brown rice with bran has more food value than white rice. To get white rice, the bran is taken off and the rice is polished.

There are two main types of rice. The first is *long-grained* rice. When this rice is cooked the kernels are fluffy. The second kind is *short-grained* rice. When this rice is cooked the kernels tend to stick together in lumps.

All types of rice swell when they are cooked in water. The hard kernels absorb the water. Usually one cup of rice will make three cups of cooked rice.

Have you ever eaten *wild rice?* Did you know that wild rice is not really a rice? Wild rice is a tall grass. Wild rice was used for food mostly by American Indian tribes around the Great Lakes. Today it is still gathered and sent to market. Wild rice is more expensive than white or brown rice.

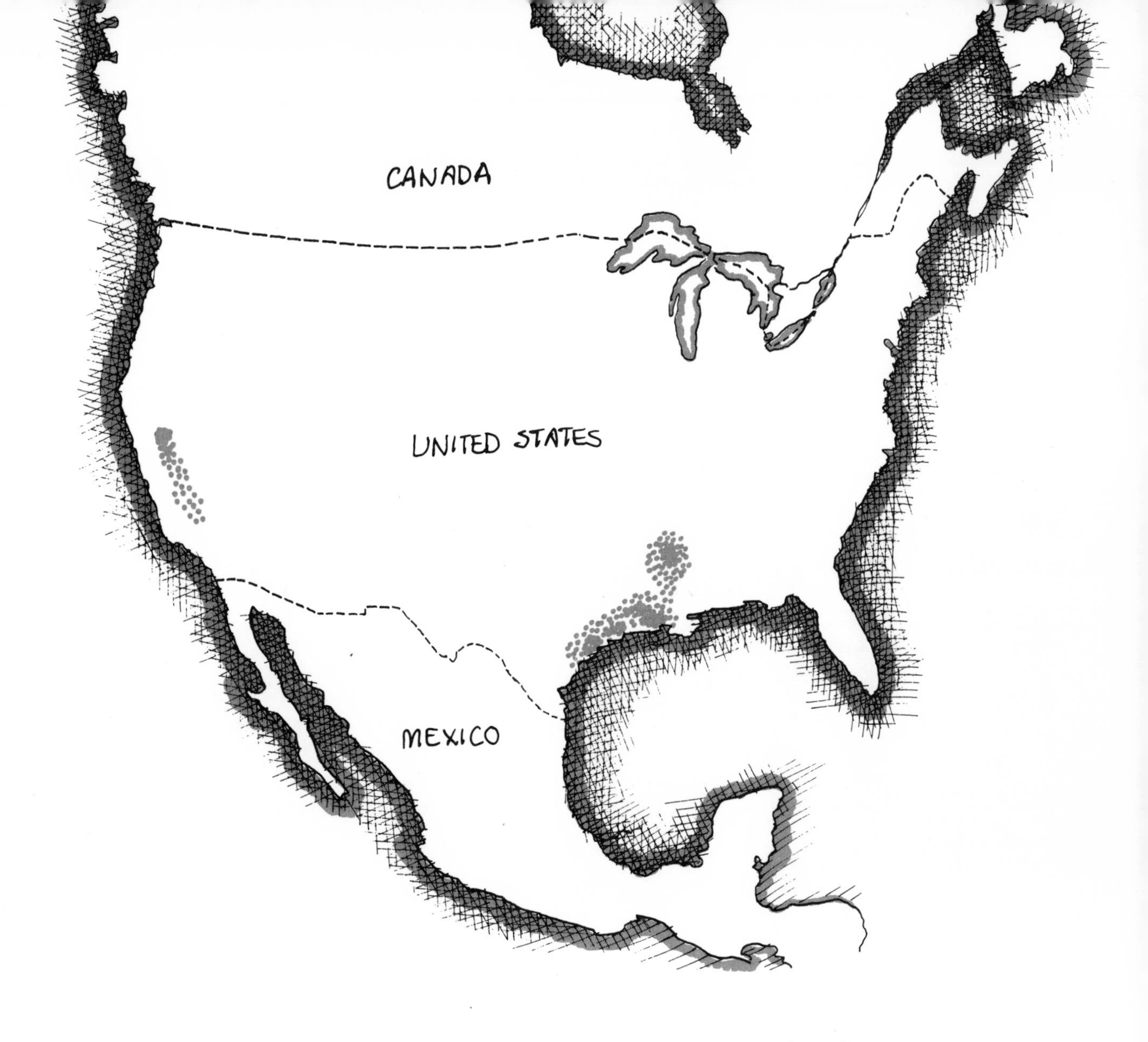


Most of the true rice that we eat in the United States is grown in hot southern states. Texas, Arkansas, California, and Louisiana grow the most rice in America. Rice is also grown in Missouri, South Carolina, Tennessee, and Mississippi.

Rice growing in America happened by chance. In 1694 a ship was sailing to England from Madagascar, an island in the Indian Ocean. One day the ship was caught in a storm. It had to stop at Charleston, South Carolina for repairs. Landgrave Thomas Smith, the governor of the colony, happened to go on board. He saw rice seeds in the cargo. Smith asked the captain if he could have some. The captain gave him some rice to thank him for his help.

Farmers had already tried to grow rice in America, but it would not grow well. Smith planted his rice seeds in some swampy soil in South Carolina. At the end of the season he had such a good crop that he could feed almost everyone in the colony.

*Planes replace workers in American rice fields.*

Rice Council of America

Today millions of tons of rice are grown in the United States each year. American farmers can grow that much because they use very modern tools and machines. Farmers in America don't grow rice in the same way that Asian farmers do. A lot of the work is done by people in airplanes instead of people in the fields.

In the United States the rice seeds are usually soaked in water for 48 hours. Then they are dropped from airplanes that fly over the rice fields. Next, water is piped in to flood the fields.

When the shoots sprout, fertilizer is spread from airplanes onto the fields. No one has to pull up the shoots and replant them in rows so they can be weeded easily. Airplanes just spread weed killer over the fields.

American farmers don't cut rice or thresh it by hand. Huge machines called *combines* are driven into the fields. They cut, gather, and thresh the rice all in one trip.

Sperry/New Holland Photo

Then the grain is taken to mills. It is dried and polished in large machines.

Scientists have been studying rice for many years. They are trying to help farmers in poor countries grow more and better rice crops. More rice is needed to feed the world's hungry people. Scientists have found that if they plant dwarf rice with long-stemmed rice, they can grow bigger crops. These crops also ripen faster than regular rice.

Growing more and better crops in underdeveloped countries is often called the Green Revolution. The Green Revolution has caused larger rice, corn, and wheat harvests and developed easier ways to grow them. As a result, more and more of the hungry people in the world will be fed.

FAO Photo by D. Mason

*New types of rice have increased rice production around the world.*

## GLOSSARY

**A Frame:** a pole some farmers use to transport rice from field to mill.

**Bran:** the broken covering of the grains of wheat, rye, etc., which is separated from the inner part that is made into flour. Bran is used as fodder and in cereal, bread, and other foods.

**Cereal plants:** any grass that produces grain which is used as a food; wheat, corn, rice, oats, and barley are cereal plants.

**Combines:** machines for harvesting and threshing grain.

**Husk:** the dry outer covering of certain seeds or fruits. An ear of corn has a husk.

**Long-grained rice:** the kind of rice that cooks into fluffy kernels.

**Monsoon:** a seasonal wind of the Indian Ocean and southern Asia. It blows from the southwest from April to October and from the northeast during the rest of the year. This wind often brings heavy rains.

**Seed beds:** muddy areas where rice seeds are first planted.

**Short-grained rice:** the kind of rice that sticks together when cooked.

**Threshing:** separating the grain or seeds from wheat, corn, or rice, etc.

**Winnowing:** when grains are sifted so that only the heavy grains are left.

INDEX

About the Author:

Franklin Watts was born in Sioux City, Iowa. He lived in the Middle West for a considerable part of his lifetime, so the subject of raising as well as eating good food comes naturally.

Mr. Watts was the founder of two publishing houses under his own name—one in the United States and one in London. As a publisher Franklin Watts specializes in books for the young and his previous titles have also been directed to young audiences. "Children are very curious creatures and it is my aim and purpose in what I write to satisfy some of their curiosity," Mr. Watts says. "In fact, I hope to increase their desire for more information. While there have been books on food for the young, most of them have started with growing or the geography of foods. Here I am starting right where the child is—the fun of eating the food—so these books are planned working back from the food in the dish to the place where it is grown."

About the Artist:

Born in India, Sam Shiromani has made Chicago his home. A dropout from the world of advertising, he devotes most of his time to free-lancing art and photography.